Commercial Structures

CARSON DUNLOP & ASSOCIATES

Dearborn™
Real Estate Education

President: Roy Lipner
Publisher and Director of Distance Learning: Evan M. Butterfield
Senior Development Editor: Laurie McGuire
Content Consultant: Alan Carson
Production Coordinator: Daniel Frey
Creative Director: Lucy Jenkins
Graphic Design: Neglia Design Inc.

Published by Dearborn™ Real Estate Education
a Division of Dearborn Financial Publishing, Inc.®
a Kaplan Professional Company®
30 South Wacker Drive
Chicago, IL 60606-7481
www.dearbornRE.com

Printed in the United States of America.
03 04 05 10 9 8 7 6 5 4 3 2 1

Table of Contents

OBJECTIVES AND OVERVIEW

OBJECTIVES

At the end of this book, you will be familiar with common commercial building structures. You will recognize the main components that make up the structure and understand common structural deficiencies. You will know enough about commercial building structures to discuss them confidently with your clients and consultants.

OVERVIEW

INSTRUCTIONS

There are four Study Sessions in this book. Each Study Session should take between 30 to 45 minutes to complete. The entire book will take you between 2.5 and 3.5 hours to complete. It is all right if it takes more or less time than we have estimated. Don't rush.

WHAT'S IN HERE?

This book covers the most common commercial building types and structural components.

The four Study Sessions cover –

Definitions and descriptions of structural components.

Structural assemblies and building types.

Typical problems and concerns.

Inspection of the structure.

'Need-to-Know Basis'

Information is presented on a 'need to know basis.' By the end of this book, you will understand the basics of commercial building structures. This will assist when dealing your client's concerns about the building and allow you to guide your client in the right direction if further investigation is required.

STUDY SESSIONS AND QUICK QUIZZES

The Sessions contain text, illustrations and photographs. At the end of each Study Session there is a Quick Quiz to help you review what you have learned. The answers to the Quick Quiz questions are given for your reference.

HERE WE GO!

STUDY SESSION 1

Structural Component Definitions and Descriptions

1. This section introduces terminology and definitions necessary for understanding commercial structures and what goes wrong with them. While you may already know many of the terms, there will likely be some you are not completely familiar with.

2. By the end of this Study Session, you should be able to –

   - explain the purpose and function of a footing.
   - explain the purpose and function of piles.
   - describe the two functions of the foundation walls.
   - describe the loads that are handled by beams and girders.
   - describe the loads on a joist.
   - describe the function of a column.
   - explain what reinforced concrete is and why it is a desirable material.
   - describe the characteristics of pre-cast concrete panels.
   - define deflection.
   - give an example of a cantilever.

3. This Study Session should take you about 40 minutes to complete.

4. Quick Quiz 1 is included at the end.

KEY WORDS

- footing
- piles
- foundation
- beam
- girder
- joist
- column
- substructure
- superstructure
- reinforced concrete
- reinforcing bar (re-bar)
- pre-cast concrete panel
- deflection
- cantilever

INTRODUCTION

The structure is a major concern for anyone responsible for a commercial building. If the structure is bad then it doesn't matter how good the HVAC system is or how much revenue the building generates.

The perfect structure will perform its intended function for many years with little maintenance. A bad structure could be so devalued to the point the building is deemed to be worth 'land value minus demolition costs.' Most structures fall somewhere in between these two extremes. While there are no perfect building structures out there, it is fairly uncommon that the structure is so bad it cannot be salvaged. The key is to know what you are dealing with.

While commercial real estate professionals should not take on the responsibility of verifying the structural integrity of a commercial building, it is important for you to know the basics so you can speak to consultants and clients in a knowledgeable fashion and understand the implications of a structural concern raised.

FOOTING

The building ***footing*** keeps the building from sinking into the soil.

Footings Keep Buildings From Sinking

A footing is a widening at the base of a foundation or column. The weight of the building is spread out over a larger area of soil. This reduces the pressure on the soil. The best analogy would be, 'snowshoes.' In deep snow, a hiker would sink into the snow, making it difficult to walk. If the hiker puts on snowshoes, his weight is spread out over a larger area, the pressure on the snow is reduced and the hiker walks on top of the snow.

Footings Are Located under Columns and Foundations

The following illustration shows a cross-section of a foundation wall and a column with footings. Note: The concrete floor is slab-on-grade; therefore, the floor does not support the weight of the building in any way. The column penetrates the concrete floor and rests on the footing beneath. The weight of the building superstructure is exerted on the columns and the foundation wall. The load is transferred to the footings. The footing spreads the weight out over a larger area. The load is then exerted on the soil.

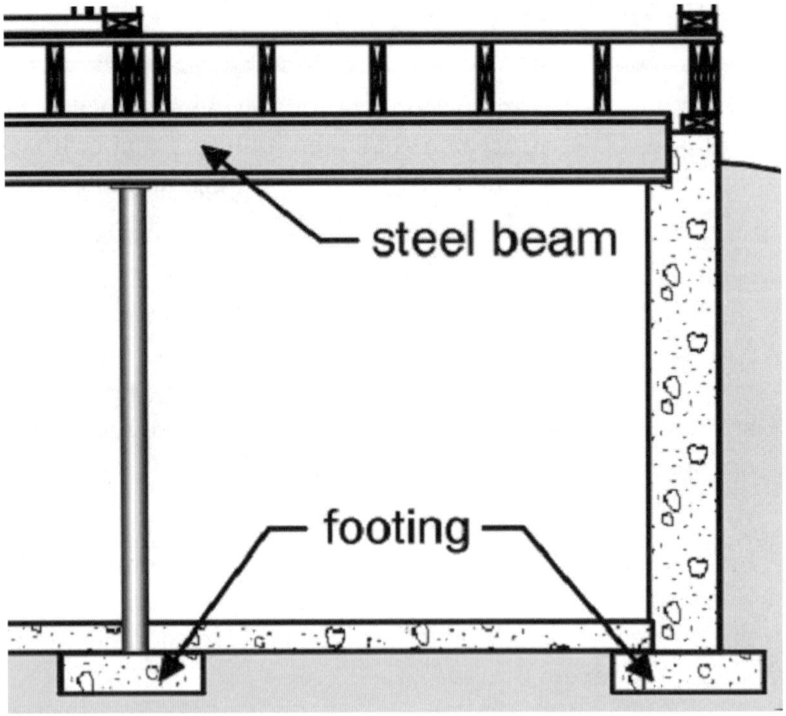

Larger Footings for Heavier Buildings or Weaker Soil

The width of the footing depends on the weight of the building and the strength of the soil. A heavier building would tend to need larger footings for a give type of soil. Also, larger footings would tend to be required for a weaker soil. Conversely, smaller footing would be required for stronger soils. In fact if the soil is so strong, like solid rock for example, no footing may be required at all.

PILES

Piles Used for Weak Soil

If the soil is so weak that even a very wide footing will not support the building, the building can be supported on ***piles***.

A pile is like a long column hammered into the ground until it hits solid rock or until the friction around the pile is so great it cannot be hammered any deeper. The strength of the soil supporting the pile can be estimated by the amount of force applied to the pile with no movement of the pile.

The following illustration shows piles that have been hammered down to solid rock. The building is supported on the piles rather than the soil.

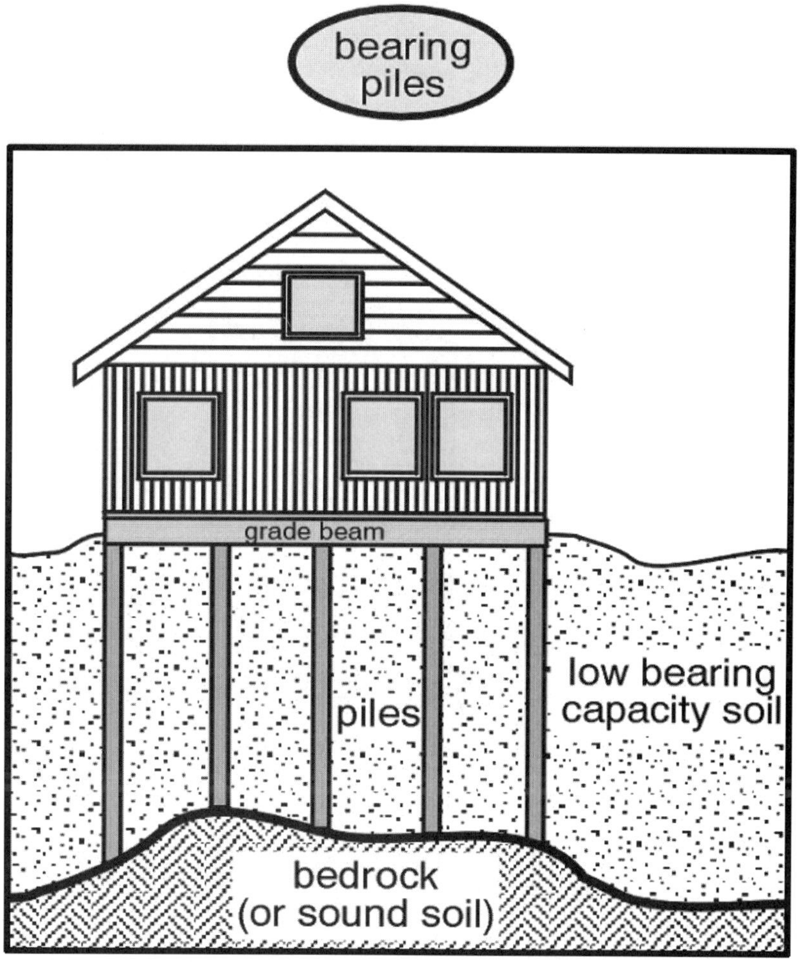

piles bearing on bedrock or sound soil below

The illustration below shows piles hammered into the ground until the friction is so great they cannot be hammered any more.

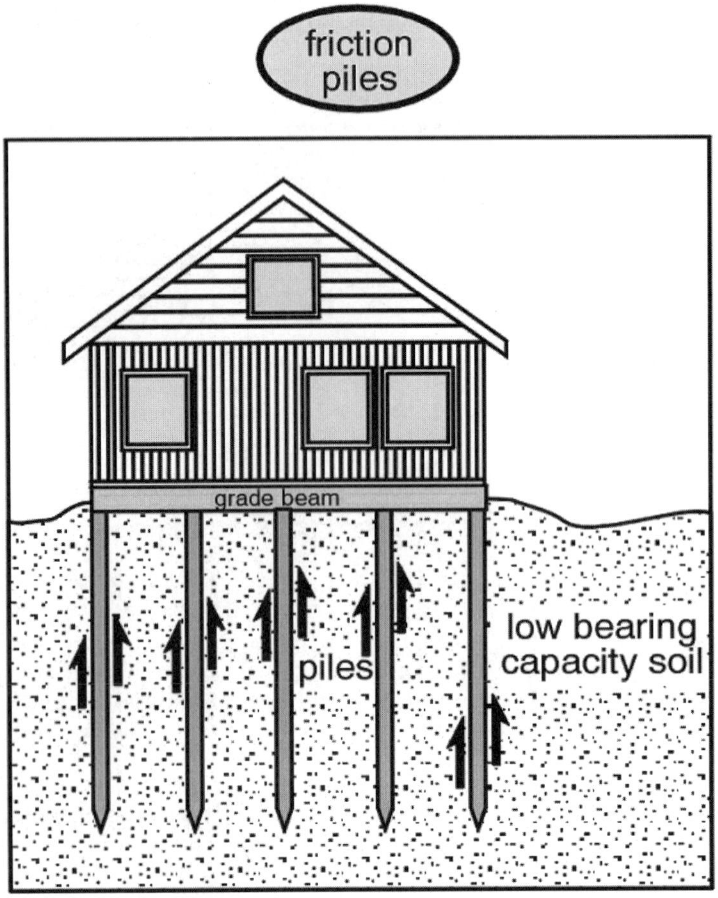

piles providing support through skin friction

FOUNDATION

Foundations rest on footings. For the most part, foundations are made of concrete. If the building has a basement or a crawlspace, the foundation wall forms the walls of the basement area. The foundation wall has two functions:

1. support the weight of the building and
2. to act as a retaining wall to hold back the surrounding soil when there is a basement or underground parking garage.

SUPPORT THE WEIGHT OF THE BUILDING

Foundations Support the Building Superstructure

The foundation wall supports the weight of the building. Another way to look at it is that the foundation wall transfers the weight of the building to the footings, which in turn transfer the weight of the building to the soil. The soil ultimately supports the building.

Examples of loads that are supported by the foundation walls are –

the weight of the building walls;

the roof loads (e.g. snow); and

the weight of the floor structure and the people and things on the floor.

The following illustration shows foundation walls supporting building loads. On the left, we see floor loads being supported on a foundation wall. On the right we see the wall of the building and roof loads added to the weight of the floor.

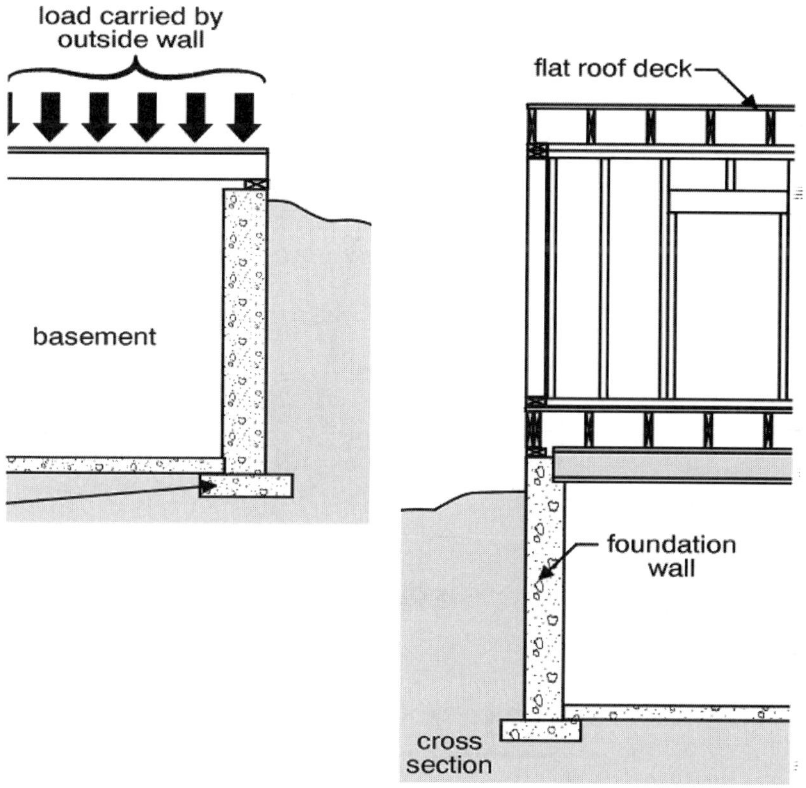

HOLD BACK THE SURROUNDING SOIL

Foundation Walls Are Retaining Walls

If there is a basement, crawlspace or underground parking garage, the soil surrounding the foundation wall exerts a great weight on the foundation wall. It is the foundation wall's job to hold back this soil. The foundation wall is a retaining wall.

The following illustration shows a foundation wall not doing a very good job of holding back the weight of the surrounding soil. This foundation wall is failing.

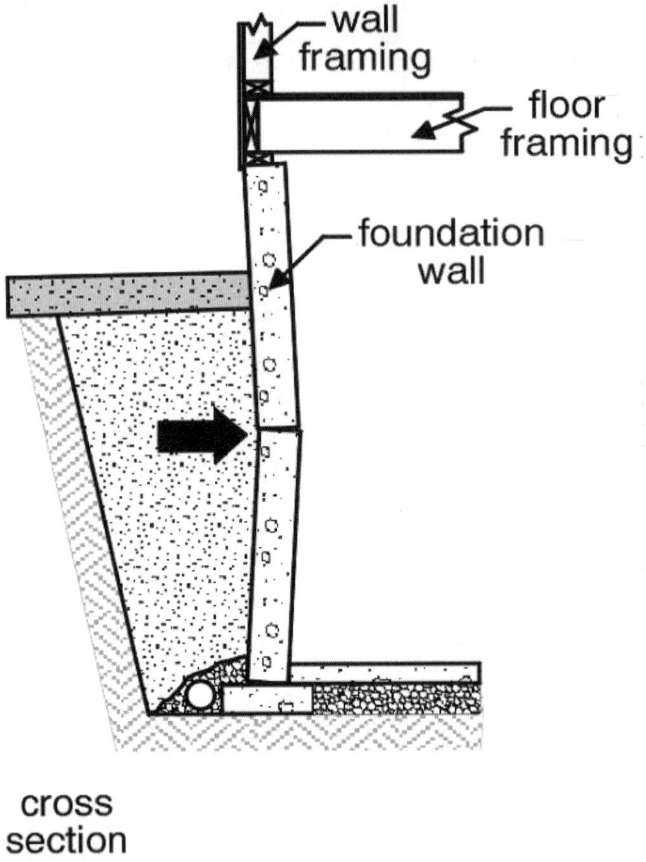

BEAMS AND GIRDERS

A ***beam*** is the same thing as a ***girder***.

Beams are large horizontal members that transfer floor loads, walls and roof loads to the columns and foundation walls.

Loads on Beams Are Vertical

The loads on beams are primarily vertical. The following illustration shows a beam with vertical loads from floor joists. The loads are transferred to two columns.

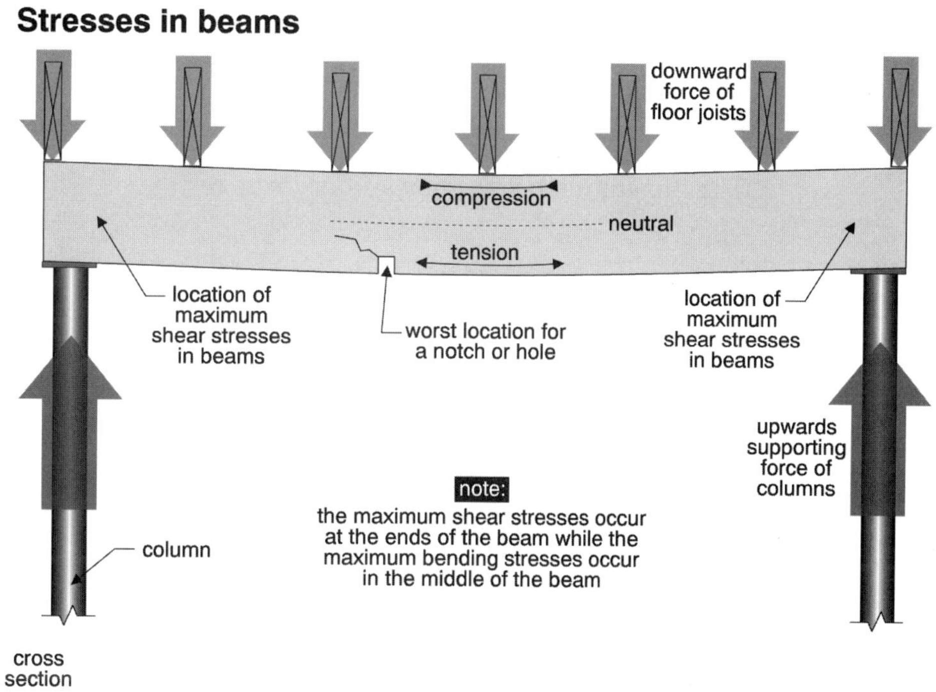

JOIST

Joists are like small beams. They transfer floor or roof loads to beams and walls.

Joists Are Like Small Beams

Joists can be made of wood, but more commonly are made of steel in commercial buildings. The steel joists have horizontal members at the top and bottom referred to as chords. Between the top and bottom chords are the web members. These steel joists are often referred to as open web steel joists (OWSJ) and will be discussed in greater detail in Session 2.

The illustration below shows floor joists resting on a main beam.

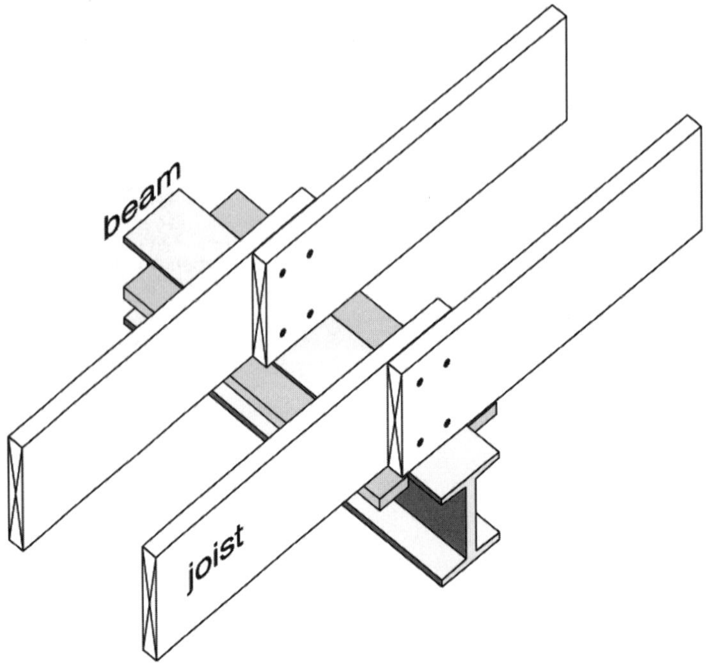

COLUMN

Columns Transfer Loads to Footings

Columns transfer loads vertically from beams to footings or to other beams. The illustration below (left) shows floor joists resting on a beam and a beam resting on a column. The column sits on a footing beneath the concrete floor slab (inset on bottom right)

Column connections

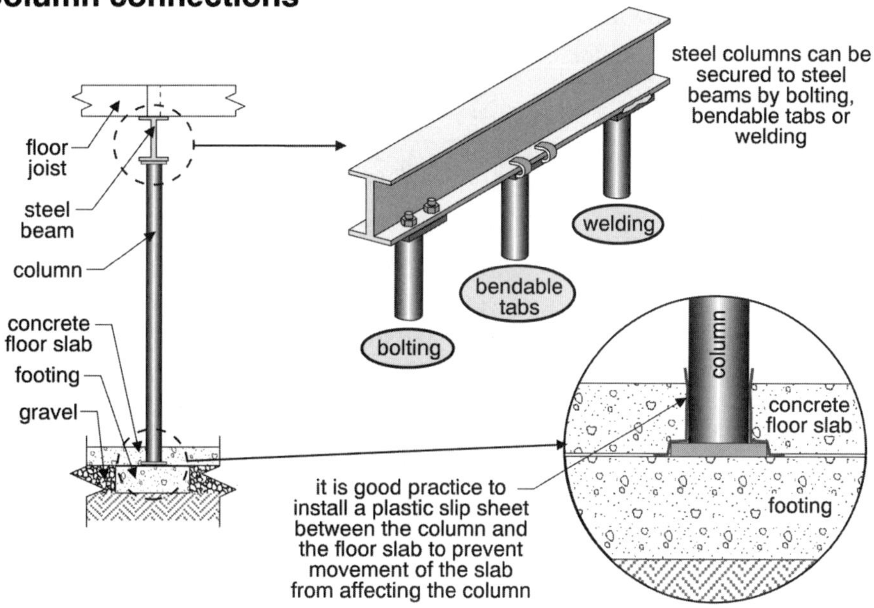

SUBSTRUCTURE AND SUPERSTRUCTURE

The ***substructure*** includes the portion of the building extending below the ground. It includes the footings and foundations.

The ***superstructure*** is the portion of the building structure starting above grade and up. In general, this is every thing from the top of the foundation and up.

REINFORCED CONCRETE AND RE-BAR

Reinforced concrete is a composite material usually consisting of concrete and steel bars (re-bar). The structural properties of the resulting composite are superior to either of the two materials alone.

Steel Bars Embedded in Concrete

Concrete is an excellent building material except it is very weak in tension (when you try to stretch it). Steel bars, on the other hand, are an excellent choice for tensile loads. The obvious answer here is to embed steel bars into the concrete in areas we expect will be under a tensile load. The steel bars will do the job in the tensile areas and the concrete will do the job elsewhere. We end up with an excellent composite called ***reinforced concrete***.

The following photograph shows the steel bars exposed underneath a concrete slab. These bars are not supposed to be exposed and are only exposed here because the concrete has been badly damage by water.

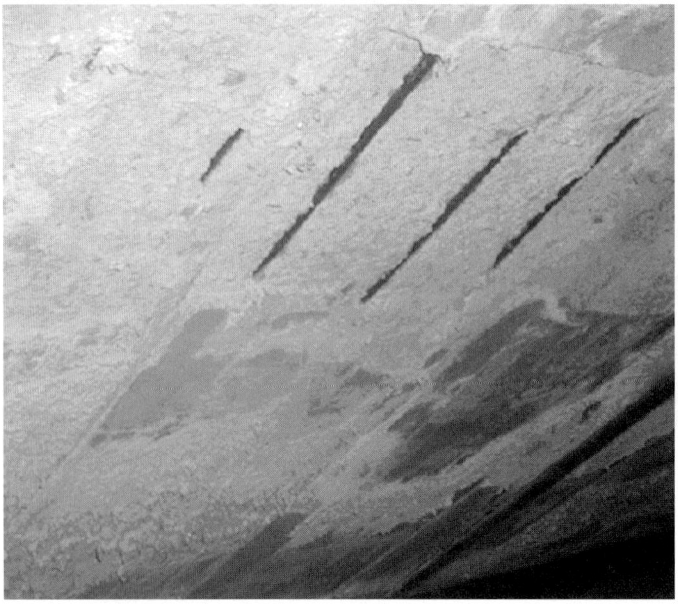

PRE CAST CONCRETE PANELS

Panels Are Hollow

Pre-cast concrete panels are reinforced concrete building units manufactured in a shop and delivered, ready to use, to the building site. They make more efficient use of the concrete because they have a hollow core. Steel bars are embedded where tensile forces are expected.

These panels can be used for floor structures or roof structures.

The following photograph shows an end view (cross section) of a pre-cast concrete panel. You can see the hollow core and the embedded steel bars (see arrow). This panel is being used for a floor structure. We are seeing the end that rests on top of a wall.

DEFLECTION

Deflection Is the Amount of Sag or Bending under a Load

For our purposes, deflection is the amount of sag in a joist, beam when a load is applied. For example, if you stand on the end of a diving board it bends under your weight. This bending is '***deflection.***'

Structural components don't bend or ***deflect*** as mush as a diving board but they do bend and sag under the loads applied, even if we don't see it. For example, if you and a few friends walk into a room in your home, the floor may sag 1/16 of an inch or more.

Deflection is Normal

Deflection is normal. Floor structures are designed to deflect a little. The illustration below shows the normal maximum allowable deflection for a floor structure with a drywall ceiling below

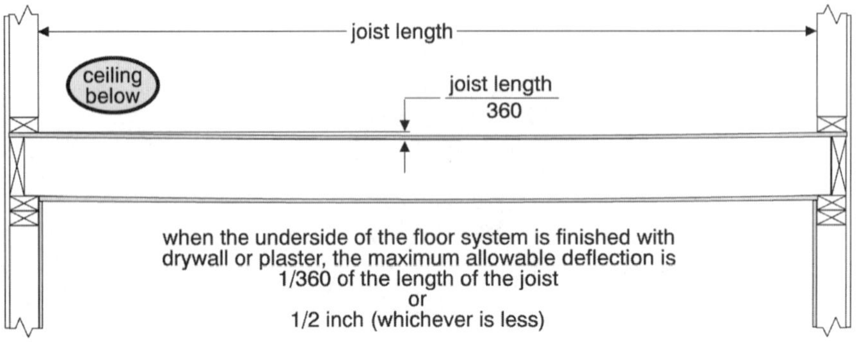

CANTILEVER

A Cantilever is Like a Diving Board

A ***cantilever*** is like a diving board. One end of the diving board is rigidly secure; the other end is free to move. An example of a cantilevered structural component is a balcony. Large apartment buildings may have concrete balconies projecting straight out from the building. The balcony is rigidly supported at the building end but at the other end the balcony is not attached to anything.

The illustration below shows a wood balcony cantilevered from a residence.

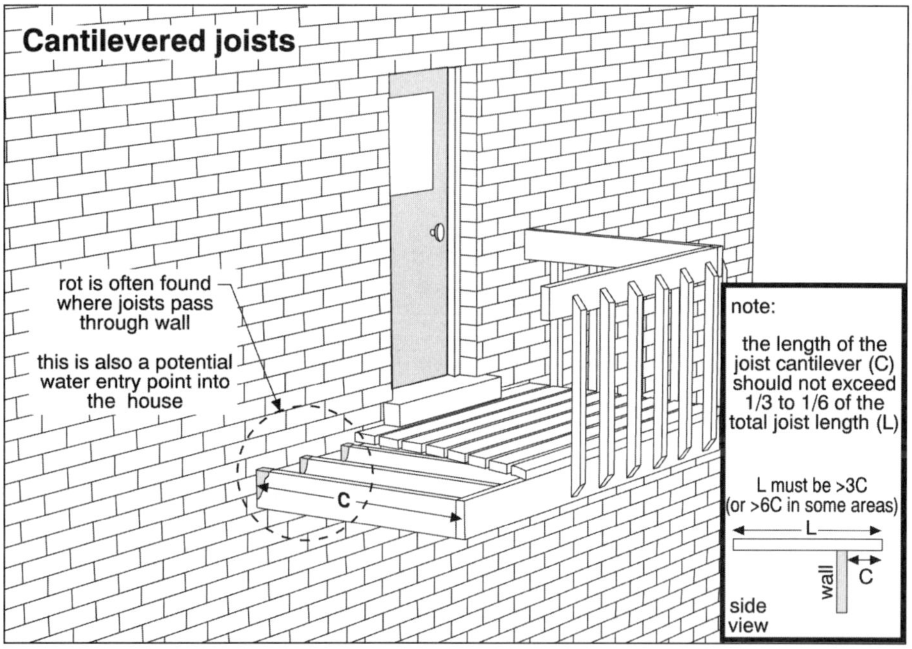

QUICK QUIZ 1

INSTRUCTIONS

You should finish Study Session 1 before doing this Quiz.

Write your answers in the spaces provided.

Check your answers against ours at the end of this book after completing the entire Quiz.

If you have trouble with the Quiz, reread the Study Session and try the Quiz again.

If you did well, it's time for Study Session 2.

1. What is the purpose of a footing?

   ____

   ____

   ____

2. What are the two functions of the foundation wall?

   ____

   ____

3. Beams transfer loads to (list two components) -

   ____

   ____

4. What do columns rest on?

   ____

5. What is reinforced concrete typically reinforced with?

   ____

6. Pre-cast concrete panels are used for (list two areas) -

   ____

   ____

7. What is a cantilever?

   ____

   ____

   ____

If you had no trouble with the Quiz, you are ready for Study Session 2

STUDY SESSION 2

Structural Assemblies and Building Types

1. You should have finished Study Session 1 and Quick Quiz 1 before starting this study session.
2. This section deals with various structural components and building types.
3. At the end of this Study Session, you will be able to –
   - describe an OWSJ.
   - describe curtain wall cladding.
   - list two differences between commercial building construction and residential building construction.
   - describe the 'tilt-up' building construction technique.
   - list two locations that a corrugated metal pan may be used.
4. This Study Session should take about 45 minutes to complete
5. A Quick Quiz is included at the end.

KEY WORDS

- open web steel joist (OWSJ)
- cladding
- spandrel area
- tilt-up building
- corrugated metal pan

SINGLE-FAMILY RESIDENTIAL VERSUS COMMERCIAL STRUCTURES

There are many differences between the structure for a single-family residential home and the structure for a commercial building. The most notable differences include the following:

Design Is More Stringent for Commercial Buildings

Commercial buildings are likely designed by an Engineer, whereas residential homes are not.

The design criteria for a commercial building are much more stringent than residential homes. For example, the floor structure of a commercial building will have been designed for a live load of 50 to 100 pounds per square foot whereas the floors of residential homes are designed for a live load of 30 or 40 pounds per square foot.

Commercial buildings typically use concrete and steel whereas residential homes use primarily wood framing.

CONCRETE STRUCTURES VERSUS STEEL STRUCTURES

Most commercial buildings are built from concrete or steel or both. In some areas, steel is the predominant structural component and in other areas concrete is the predominant structural component.

CANADA

In Canada, most large buildings are concrete structures. The columns are concrete and the floor slabs are concrete.

The small buildings such as single story industrial buildings are mostly steel structures.

USA

In the United States, most large buildings are steel structures. The columns and beams are steel.

Concrete Structures

Concrete Balconies Are Extensions of the Floor Slab

The following photograph shows a large concrete apartment building. We can see the edges of the concrete floor slabs. Follow the line made by the concrete balcony. The concrete balconies are just extensions of the concrete floor slab. The columns, beams and shear walls that form the skeleton of the building are also concrete although we cannot see them in this photograph.

Here is another concrete structure. This one is under construction. When it is finished, it will be difficult to tell if it is concrete or steel when viewed from the exterior because there are no balconies. We can see the concrete floor structure and concrete columns and shear walls. You can see the 'swing stages' in the middle of the photograph where insulation and brick veneer is being applied.

Steel Structures

The following photograph shows a steel structure being built. You can see steel columns and beams. The floor is corrugated steel pans. Concrete will be poured into the steel pans.

'TILT UP' BUILDING

Walls Are Formed on the Ground and 'Tilted Up'

The 'tilt-up' building is a method of construction rather than a building type. This construction technique is used in some areas and not in others. It is used primarily for single story industrial buildings. The technique is as follows:

Forms are carved into the earth next to where the building will stand. Concrete is poured into the forms. When the concrete has cured, it is 'tilted up' into a vertical position to make a wall of the building. You can often see the texture of the earth etched into the walls when they are lifted from their forms.

CLADDING

Cladding is the exterior surface of the building. There are many different types of cladding such as –

- masonry
- stone
- granite
- marble
- pre-cast concrete
- metal panel
- glazing panel
- stucco

Most of these systems will likely be familiar to you. A category of cladding particular to commercial buildings that deserves more explanation is ***curtain wall*** cladding.

Curtain Walls 'Hang' from the Building Frame

A curtain wall is a general term used to describe a cladding system that is entirely supported by the structural frame, but is not load bearing. Another way of putting it is that the building frame or skeleton is built and the cladding is 'hung' onto the building. The cladding system is not self-supporting.

For example, a concrete structure could be clad with brick veneer supported by the slab edges, or glazing panels hung on the slab edges and mullions. Since each brick is supported by the one below, the brick veneer is not a curtain wall. Since the glazing panels do not support each other, it is a curtain wall.

The following two photographs show how a curtain wall system is constructed.

Glazing Curtain Wall

The preceding photograph shows a steel structure. The exterior cladding is a pre-cast concrete ***curtain wall system***. The concrete panels are just hanging from the steel structure. You can see two concrete panels already in place (upper left) and another one ready to be lifted into place (bottom right). The space in between the two panels already hanging on the building is for ***glazing curtain wall***. Glazing curtain wall is also a curtain wall panel but it is made of glass.

Spandrel Area

When discussing cladding systems you may hear the term ***spandrel area***. Spandrel refers to the wall area between the top of a window on one floor level and the sill of a window on the floor level above. Essentially, the cladding system in the spandrel area must be opaque, so that the building structure and mechanical systems between floors are not visible from the exterior.

The following photograph shows how the curtain wall panels are hung on the building. In the middle of the photograph, you see the steel framing of the building photographed from inside the building. On the right half of the picture, you are seeing the back of a pre-cast concrete curtain wall panel.

The two pieces of hardware with the bolts sticking through are holding the concrete panel in place –

The top piece of hardware is for holding the panel ***in*** and adjusting how it sits relative to the plane of the exterior wall. In other words, if you look at the finished building from the street and it looks like a corner of this panel is sticking out, this bracket is not adjusted properly.

The piece of hardware on the bottom is for holding the concrete panel ***up*** and adjusting the panel so it is level. In other words, if you look at the finished building from the street and the concrete panel looks askew, then this bracket is not adjusted properly.

There will be a pair of brackets like this at each of the four corners of the panel. This is all that holds the panels in place. The panels do not rest directly on the building in any areas.

OPEN-WEB STEEL JOIST

Open-web steel joists are one of the most common components used for commercial building structures. Open-web steel joists support roof loads or floor loads and transfer these loads to beams and columns or supporting walls.

OWSJ Is Made of Chords and Webs

Open-web steel joist is abbreviated OWSJ. There is heavy steel across the top and bottom (called the top and bottom chords) with thin zigzagging steel bars (web) between the top and bottom chords. The joist is called 'open-web' because you can see through the 'web' space (the space between the top and bottom chord).

The photograph above shows open-web steel joists supporting a corrugated metal pan. The metal pan is the base of a concrete slab.

You can see the top and bottom chord of the steel joist and the zigzagging open-web between the chords.

OPEN-WEB STEEL JOISTS AND CORRUGATED METAL PAN

The open-web steel joist we saw above and the corrugated metal pan are two components of a common floor structure for commercial buildings. It is worth a look at how this system works:

The building frame (beams and columns) is assembled.

Open-web steel joists span between the beams.

A corrugated metal pan is attached (welded) to the top of the joists.

The metal pan is filled with concrete for a floor structure. The little bumps in the metal are designed to 'bite' into the concrete. The pan is a form, but also becomes part of the assembly, as the tension member.

The metal pan for a roof structure is typically smooth as concrete is not required.

The following photograph shows a corrugated metal pan ready for a pour of concrete to make a floor structure. Note the steel lip at the edge acts as the form to contain the concrete when its poured.

QUICK QUIZ 2

INSTRUCTIONS

You should finish Study Session 2 before doing this Quiz.

Write your answers in the spaces provided.

Check your answers against ours at the end of this book after completing the entire Quiz.

If you have trouble with the Quiz, reread the Study Session and try the Quiz again.

If you did well, it's time for Study Session 3.

1. Give two differences between the structure of a commercial building and a single-family residential building -

   ____

   ____

   ____

   ____

2. What is a curtain wall?

   ____

   ____

   ____

3. What is the spandrel area?

   ____

   ____

   ____

4. The zigzagging steel bars between the chords of the open-web steel joist are called the -

   ____

5. What is a 'tilt-up' building?

____

____

____

6. What is an OWSJ?

____

7. Where might a corrugated steel pan be used?

____

If you had no trouble with the Quiz, you are ready for Study Session 3

STUDY SESSION 3

1. You should have finished Study Session 2 and Quick Quiz 2 before starting this study session.
2. This section deals with some of the common problems that occur with commercial building structures.
3. At the end of this Study Session, you will be able to –
    - describe differential settlement.
    - describe uniform settlement.
    - describe tipping settlement.
    - explain the significance of a horizontal crack in the foundation.
    - explain how a pre-stressed suspended concrete floor slab is constructed.
    - explain why suspended floor slabs in parking garages are a potential problem.
    - give three examples of hostile environments that can affect the structure of the building.
    - explain a common problem with concrete balconies.
4. This Study Session should take about 45 minutes to complete
5. A Quick Quiz is included at the end.

KEY WORDS

- uniform settlement
- differential settlement
- tipping settlement
- foundation failure
- suspended slab
- pre-stressed
- pre-tension
- post-tension
- membrane
- catastrophic failure
- buckling
- column
- hostile environment
- concrete balcony

BUILDING SETTLEMENT

Settlement is Common

Building settlement occurs in almost all buildings, to varying degrees. The implication of the settlement can range from completely insignificant to a major problem.

There are three basic types of settlement:

1. uniform settlement
2. tipping settlement
3. differential settlement

Types of settlement

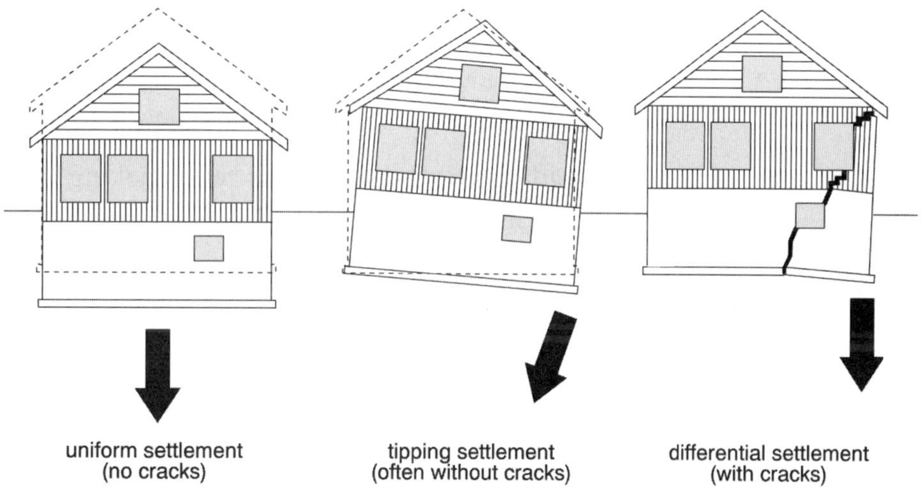

UNIFORM SETTLEMENT

Building Moves Straight Down

Uniform settlement means the building has moved straight down. It is often very difficult to determine that this has occurred because the building is still standing straight and there are no cracks in the exterior.

The implications of uniform settlement are –

- The building may continue to settle.
- The services to the building may be affected. For example, the main drain may no longer slope toward the street or gas piping or underground electrical service conduit may bend and sever.

TIPPING SETTLEMENT

No Cracks

Tipping settlement means the building tips as a unit, usually with no cracks in the exterior of the building.

The implications of tipping settlement are –

- The building may continue to settle.
- The floors may not be level, affecting building use.
- Tipping settlement is most often associated with pockets of poor soil or soil that is eroding.
- Substantial costs may be associated with stabilizing the building.

DIFFERENTIAL SETTLEMENT

Cracks

Differential settlement means one part of the building has settled while the rest of the building has not. This is the most common type of settlement.

Differential settlement results in settlement cracking in the foundation and exterior walls of the building.

The implications of differential settlement –

- The building may continue to settle.
- The floors may not be level, affecting building use.
- Differential settlement is most often associated with pockets of poor soil or soil that is eroding.
- Substantial costs may be associated with stabilizing the building.

FOUNDATION FAILURE

While differential settlement results in cracks in the foundation, it is not usually a significant foundation failure. There may be a performance issue if the building continues to settle, but the ability of the foundation to hold up the building and hold back the soil will not be appreciably affected.

Horizontal Cracks

We mentioned in Study Session 1 that the foundation has two functions, to hold up the building and to hold back the surrounding soil. The main type of foundation failure occurs when the foundation can no longer hold back the surrounding soil. The foundation develops horizontal cracks near the mid-height. If the foundation is not reinforced at this point, the foundation will bulge and eventually fail completely (collapse).

The illustration following shows a typical horizontal crack in a building foundation, indicating foundation failure.

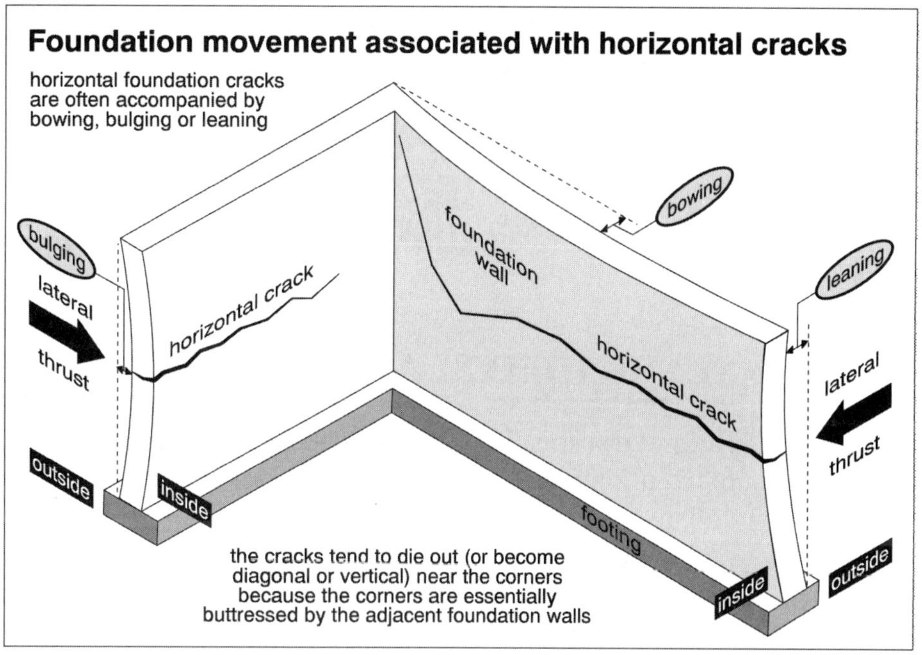

The following photograph shows the basement area of a commercial building. There is a horizontal crack running the length of the wall. The inset to the right is a close-up of the crack. This crack is about $1/8^{th}$ of an inch wide. This is indicating foundation failure. Something will have to be done to prevent this from progressing further.

SUSPENDED SLAB FAILURE

Parking Garage Failures

Failure of suspended concrete slabs is an issue that has grabbed a great deal of attention lately because of a series of near catastrophes, including the collapse of a parking garage at a shopping mall in Canada. In fact, it is parking garages that are experiencing the problem. Let's look at the problem more closely:

SUSPENDED CONCRETE SLABS

Pre-stressed Concrete Slab

A good example of a suspended concrete slab is a level of a multi-level parking garage. The concrete slab spans between columns and walls. The concrete is reinforced with steel. Conventionally, the steel reinforcing was steel bars embedded in the concrete ('re-bar'). More recently, a technique was developed to reinforce the concrete with steel cables rather than steel bars. This technique is called pre-stressed concrete slab construction. The benefit of this technique is that a thinner slab can be used. This means less concrete (less costly) and more headroom.

Prestressed and post-tensioned concrete floor slabs

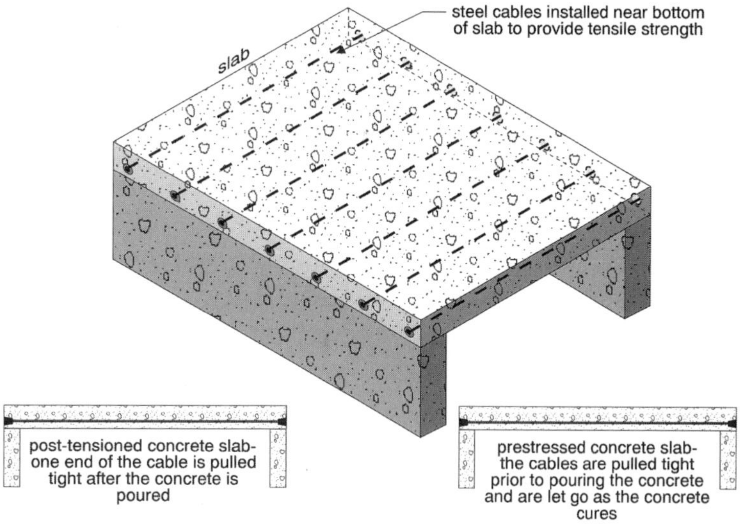

The illustration above shows the pre-stressing technique. Cables are embedded in the concrete near the bottom of the slab. When the concrete cures, a tension load in the cables is applied to compress the concrete. There are two methods to do this –

Pre-tension: The pre-tension technique involves –

- stretching the cables and setting into the form
- pouring concrete into the form around the stretched cables
- letting the concrete cure
- releasing the cables
- the cables compressing the concrete when they are released

Post-tension: The post-tension technique involves –

- setting the cables into a sleeve
- setting the sleeve with the cable into the form
- pouring the concrete around the sleeve
- letting the concrete cure
- pulling the cables tight and anchoring them at either end of the slab.

Both the pre-tension and the post-tension technique achieve the same end result – tightly stretched cables that in turn squeeze the concrete.

THE PROBLEM

The pre-stressed concrete slab is a technique that is generally reliable, but the following problems have been encountered.

Catastrophic Failure

In a post-tensioned concrete slab, a cable is stretched extremely tightly and anchored at either end. This cable is like a spring, ready to be released. What happens if there is a failure at the point of attachment of the cable or at any spot along the cable? The cable will release with explosive force. The release of the cable alone could kill someone. In addition, the slab itself may fail without the strength of all of the cables to support it. Instead of spreading the load out over many pieces of re-bar, we have concentrated the load in few steel cables.

The concern with pre-stressed concrete slabs is the danger of catastrophic failure. In practice, these cables do not just fail, there has to be some outside influence to cause them to fail. Unfortunately there exists just such an outside influence: water mixed with salt. This corrosive mixture comes from cars in the winter if salt is used on roads. In areas near the ocean, the salt is in the air and less of a concern.

This is why there is such a focus on parking garages. In fact, a building owner or purchaser of a building with an underground parking garage would be well advised to engage a qualified Structural Engineer to carefully inspect this type of structure. Some mortgage lenders will not take on buildings that have this type of floor structure.

A Concern for Conventionally Reinforced Slabs

The concern for salt and water also applies to conventionally reinforced concrete slabs. Although catastrophic failure is not likely to occur without ample warning, the corrosion of re-bar in the slab is a concern from a monetary point of view due to the high repair costs. As the steel re-bar corrodes, it expands the expansion of the steel causes the concrete to crack, accelerating the deterioration and reducing the effectiveness of the steel.

In conventionally reinforced slabs, it has become common to use epoxy-coated steel to protect it from salt and water.

MEMBRANES

Suspended Slabs in Parking Garages Need Membranes

Because of this problem with corrosion of reinforcing steel, all suspended-slab parking garages subjected to salt and water must have a protective membrane on the surface to protect the concrete and the steel reinforcing within. Existing slabs without such a membrane should have a membrane installed. Sometimes a concrete sealer is used as a less expensive alternative to a membrane.

The following photograph shows an underground parking garage with a membrane on the surface. You can see the membrane as it extends about six inches up the 'shear-wall' in the centre of the photo.

BUCKLING OR DAMAGED COLUMN

Columns and Forklifts Don't Mix

A common problem in industrial buildings is damaged columns. Usually the culprit is the forklift. With pallets being moved around the warehouse on a constant basis, it is inevitable that a forklift will collide with a column. If the damage is minor, reinforcing or no repairs at all may be required. If the damage is major, the column will have to be replaced.

The following photograph shows a large column damaged by a forklift. The damage is significant and the column will likely have to be replaced.

The following photograph shows a column damaged by a forklift. In fact the fork penetrated through the base of the column. This is minor damage and can likely be repaired.

HOSTILE ENVIRONMENT

Chemicals and Moisture May Damage the Building

Very often the process conducted in an industrial building can affect the structure. Any process generating a great deal of moisture or producing corrosive chemicals may damage the building. The following are examples of buildings that may be at risk:

- arenas
- metal plating facility
- dry cleaning facilities
- frozen food storage
- any high humidity environment

The following image shows two photographs, background and inset. They both show OWSJ supporting metal pan floor deck and roof deck above. The background photo is the floor in a fairly new facility. There is a corrosive environment in this facility and we can see patches of corrosion on the underside do the floor decking. The inset photograph shows OWSJ and roof decking in a metal plating facility. Every inch of steel in this photo is covered in corrosion.

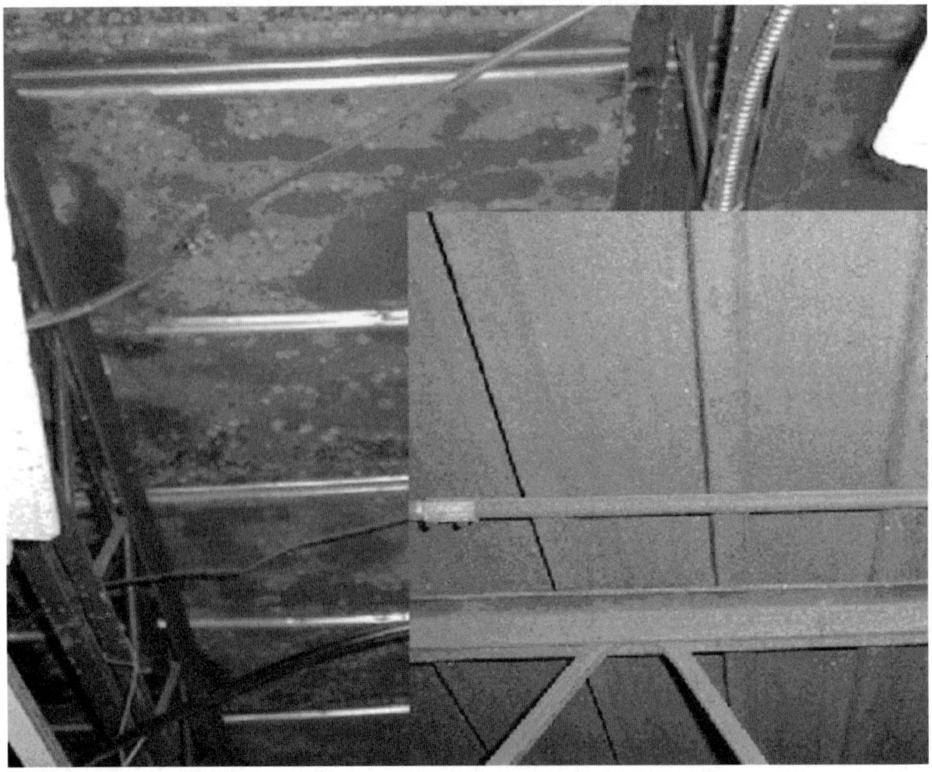

BALCONY DAMAGE

Steel Expands As It Corrodes

Concrete balconies on apartment buildings are a concern today. The concern is that the steel railings embedded in the concrete are corroding. The corroded steel expands and cracks the concrete. The concrete then falls off.

In more severe cases, the steel reinforcing embedded in the concrete is affected and the balcony structure is in compromised.

The following photograph shows an apartment building undergoing repairs to the balconies. You can see the swing stages for the workers and support posts on the balconies. In this case, only localized repairs are being done. This is an extremely expensive undertaking because all work is done painstakingly from a swing stage. Virtually every balcony is affected.

Each localized repair involves –

- chipping away loose concrete;
- painting or replacing the re-bar;
- making a concrete form for the affected area;
- pouring concrete into the form; and
- finishing / replacing railings, etc.

Concrete balconies in every major city are being repaired. This is very expensive work.

The following photograph shows a smaller apartment building. Every other balcony is being replaced. The concrete is chipped away until all that is left is the steel. The steel is repaired and primed and painted as needed and a new concrete balcony is poured. The balcony below is used to support the forms for the concrete above (you may be able to make this out at the back of the building on the right). Once every other balcony is done, the new balconies are used to support the repairs for the remaining balconies.

QUICK QUIZ 3

INSTRUCTIONS

You should finish Study Session 3 before doing this Quiz.

Write your answers in the spaces provided.

Check your answers against ours at the end of this book after completing the entire Quiz.

If you have trouble with the Quiz, reread the Study Session and try the Quiz again.

If you did well, it's time for Study Session 4.

1. What is the significance of uniform settlement?

    ____

    ____

    ____

    ____

2. What does a horizontal crack in a building foundation indicated?

    ____

    ____

3. What is the enemy of a suspended concrete slab in a parking garage?

    ____

    ____

4. What is the purpose of the membrane on a suspended concrete slab in a parking garage?

    ____

    ____

5. What is the most common cause of damage to columns in a warehouse?

   ____

6. List three examples of hostile environments that could affect the building structure -

   ____

   ____

   ____

7. What is the concern with concrete balconies?

   ____

   ____

If you had no trouble with the Quiz, you are ready for Study Session 4

STUDY SESSION 4

1. You should have finished Study Session 3 and Quick Quiz 3 before starting this study session.
2. This section describes the investigations that can be done on the structure and what kind of information they provide.
3. At the end of this Study Session, you will be able to –
   - explain the difference between a performance-based structural review and a design analysis.
   - list two nondestructive concrete testing techniques.
   - list two destructive concrete testing techniques.
   - explain what information can be obtained from a soil test.
4. This Study Session should take about 30 minutes to complete
5. A Quick Quiz is included at the end.

KEY WORDS

- performance-based review
- design analysis
- nondestructive testing
- destructive testing
- chain drag test
- Schmidt hammer test
- half-cell testing
- core sample
- chloride ion testing
- compaction

PERFORMANCE-BASED REVIEW VERSUS DESIGN ANALYSIS

There are two philosophies when it comes to assessing the building structure:

1. the ***performance-based review*** and
2. the ***design analysis***.

Of course the best way to review the structure is to be there when it is being built. Since that is not an option, we have to do what we can.

PERFORMANCE-BASED REVIEW

You Can't See All of the Structure

The performance-based review philosophy assumes that you cannot see all of the structure. There are critical components that are concealed and may be faulty. If there is a structural deficiency, over time it will manifest itself in some way we can identify visually.

This works very well for existing buildings because they have some history. They have had a chance to settle. Most structural deficiencies will be apparent to the trained observer.

Inherent in this philosophy is the fact that some structural deficiencies will not be detected.

Typically, a visual review is carried out to assess a building structure. If the trained observer has reason to believe further investigation is required in some area, that investigation can be carried out in more detail. Of course this relies heavily on the experience and credibility of the observer. The observer has to be confident in the decision for further investigation and should be able to justify the request.

This is a little like a medical General Practitioner (GP). The GP does not arbitrarily call for complicated and potentially damaging further investigation if there is no reason to believe it is necessary.

DESIGN ANALYSIS

One area where the performance-based review does not work as well is with newer buildings. If the building is only six months old, it has not gone through a year of seasons. It has not had a chance to settle and the process inside the building has not had a chance to affect the structure.

The result is that structural deficiencies due to design, construction or the environment have not had a chance to manifest themselves into something that can be observed.

This is not to say a performance-based review is worthless; it just does not work as well with newer buildings. In this case, a design review may be necessary to assess the structure.

A design review involves having a Structural Engineer go through the 'as-built' drawings of the building and verify the design. This involves detailed calculations and is time consuming. This would be followed by a detailed review of the actual construction of the building. As you can imagine, this is very expensive.

In practice, the design analysis or design review has only limited usefulness and application.

NONDESTRUCTIVE TESTING VERSUS DESTRUCTIVE TESTING

Further Investigation

If, during a performance-based review, a concern is identified, further investigation may be required. Further investigation is useful for –

- verifying assertions made during a performance-based review;
- investigating an area prone to problems that cannot be observed visually; and
- determining the extent of a problem that has been identified and defining the scope of the repair

NONDESTRUCTIVE TESTING/INVESTIGATION

Nondestructive testing does not damage the building or any finish surfaces. Examples of nondestructive testing or investigation are –

- x-ray a weld on a structural steel connection
- x-ray concrete
- ultrasonic testing of concrete
- a medical doctor listening to your heart with a stethoscope

DESTRUCTIVE TESTING/INVESTIGATION

Destructive testing does not necessarily damage the building but there may be damage to the finish, drywall may have to be removed, a sample may have to be removed and repaired. Examples of destructive testing are –

- removing finish surfaces to get a better look at a structural component;
- taking a core sample from the concrete to test its compressive strength, chloride ion (salt) content, air entrainment, etc.;
- taking a sample of soil to verify bearing capacity of the soil; or
- a medical doctor doing a biopsy on a suspected tumor.

CONCRETE SLAB TESTING

As an example of applying non-destructive and destructive testing techniques, let's consider a suspended concrete floor slab in a parking garage. Both nondestructive testing and destructive testing may be necessary.

NONDESTRUCTIVE TESTING

Nondestructive tests that can be done on concrete slabs are as follows:

The Chain Drag Test

While this sounds very high tech, it's actually very primitive. A person walks through the entire floor area, dragging a chain over the concrete surface. An experienced technician listens to the sound it makes. The sound of the chain 'dancing' on the concrete surface reveals areas of topside delamination.

The Schmidt Hammer Test

This test is done with a Schmidt hammer, or any hammer for that matter. The hammer is used to percuss (tap) on the concrete slab to listen for hollow areas (concrete delamination caused by poor installation or corrosion of reinforcing steel).

Half-cell testing

This nondestructive technique is used to investigate the integrity of the steel in reinforced concrete.

Electrodes are connected to the reinforcing steel. A voltage is measured. The voltage relates to its potential for corrosion.

DESTRUCTIVE TESTING

Destructive tests that can be done on concrete slabs are as follows:

Compressive strength test

This involves talking a core sample from the concrete slab and crushing it in a machine to determine its compressive strength. This may be a useful test for a technical audit or when changing the building usage and loading.

Chloride ion test

This test is used to determine the potential for corrosion of reinforcing steel in concrete. A sample of the concrete is taken and tested for chloride ion (salt) content. The higher the chloride ion content, the higher the potential for corrosion of the steel.

INVESTIGATION OF CONCRETE BALCONIES

We mentioned in Study Session 3 that concrete balconies are a big concern because of the high cost and high probability of repairs.

Investigation of balconies is primarily a performance-based review. In some cases, more detailed investigation is warranted. Most of the techniques described for parking garage slabs also apply to concrete balconies.

SOIL TESTING

In this context we are talking about testing the soil for its structural characteristics (geo-technical) not testing for contaminants (environmental).

This test is common when developing raw land, prior to the construction of the building. The test results will determine the size and type of footing required for the building.

This test may be warranted for an existing building that has settled substantially. When you see a building that has settled, you are only seeing a snapshot of one point in time. It is very difficult to determine from a single visit if the building is continuing to move. Its like looking at a picture of a race horse and trying to determine how fast it is running.

If you could come back five years later and compare the position of the building, you would have a much better chance of identifying ongoing movement of the structure. Since this is not an option, soil testing may be required.

Soil testing involves drilling down into the soil to the depth of the building footings. A sample of soil is retrieved and the soil sample is tested. This would likely be done in a few different areas to get a good representation of the soil.

Settlement of a building means the footing has moved under the building and the soil beneath the footing has moved. This may not be an issue in that as the soil compacts, its compressive strength may increase to the point of stabilization. Alternatively, this could be an ongoing condition where the soil is too weak to support the building and the building keeps sinking into the soil.

The results of the test will give you an indication of the strength of the soil and is ability to support a load. It will also give an indication of the level of compaction and whether it is likely to compact any more given the weight of the building.

QUICK QUIZ 4

INSTRUCTIONS

You should finish Study Session 4 before doing this Quiz.

Write your answers in the spaces provided.

Check your answers against ours at the end of this book after completing the entire Quiz.

If you have trouble with the Quiz, reread the Study Session and try the Quiz again.

If you did well, it's time for the Final Test.

1. Describe the 'chain drag test' -

____

____

____

2. Describe the performance-based inspection -

____

____

____

3. Give two examples of nondestructive testing and two examples of destructive testing -

____

____

____

____

____

____

4. Describe compressive strength testing of concrete and why you might want this test done -

____

____

____

5. What information do you get from soil testing?

____

____

If you had no trouble with the Quiz, you are ready for the Final Test.

ANSWERS TO QUICK QUIZES

QUICK QUIZ 1

1) The footing keeps the building from sinking into the soil
2) The foundation wall holds up the building (supports the building superstructure) and holds back the surrounding soil.
3) Beams transfer loads to foundation walls and to columns
4) Columns rest on footings
5) Reinforced concrete is reinforced with steel
6) Pre-cast concrete panels are used for roof structures and floor structures
7) A cantilever is like a diving board. One side is constrained and cannot move while the other side is free to deflect.

QUICK QUIZ 2

1) Differences between commercial building structures and single-family residential structure –
   a) Commercial building structures are designed by an Engineer; single-family residential buildings are not.
   b) Commercial buildings have stricter design criteria, for example the design load for floor structure would be 50 to 100 pounds per square foot compared to 30 to 40 pounds per square foot.
   c) Commercial buildings are primarily built from concrete and steel while residential homes are wood frame.
2) A curtain wall system is an exterior cladding system that is not load bearing and hangs from the building structure.
3) Spandrel area refers to the wall area between the top of a window on one floor level and the sill of a window on the floor level above
4) The zigzagging network of steel bars between the chords of the open-web steel joist is called the web.

5) A 'tilt-up' building is a building construction technique where the concrete walls are formed and poured directly in the dirt next to where they will ultimately be used. Once the concrete has cured, the walls are tilted up into place.

6) An OWSJ is an open-web steel joist. This is a truss type joist used very commonly in commercial building structures.

7) A corrugated steel pan would be used as part of a concrete and steel assembly in a floor structure.

QUICK QUIZ 3

1) The significance of uniform settlement is that the building may continue to settle and require expensive stabilization. In addition, even minor uniform settlement can affect the services and utilities to the building, for example the main drain may no longer slope away from the structure.

2) A horizontal crack in a building foundation indicates impending foundation failure. Technically it indicates the foundation has failed.

3) The enemy of suspended concrete slab parking garage structures is salt and water from the automobiles.

4) The purpose of the membrane on a suspended concrete slab is to keep the salt and water out of the concrete and away from the tension cables or re-bar.

5) The most common cause of damage to columns in a warehouse is forklifts.

6) Examples of environments hostile to the building structure are; arenas, high humidity processes, dry cleaning facilities, frozen food facilities and metal plating facilities. You may have thought of a few of your own.

7) The concern with concrete balconies is the possibility of a very expensive repair bill if repairs are required. It is highly probable repairs will be required at some point, particularly if the balcony has steel railings embedded into the concrete. The steel expands as it corrodes. This leads to cracking of the concrete and pieces of concrete falling off the building.

QUICK QUIZ 4

1) The chain drag test is a test to 'sound out' delaminated concrete and hollow spots in concrete. A chain is dragged along the surface and the sound of the 'rattling chain' is interpreted by an experienced person.
2) A performance-based inspection assumes that structural deficiencies will, it time, manifest themselves in something that can be seen during a visual inspection by an experienced Engineer. The inspection involves looking for evidence and clues of non-performance of the structure. This technique assumes the building has some history. If the building is less than one year old, for example, this technique is less reliable.
3) Examples of nondestructive testing and destructive testing are
   a) nondestructive – chain drag test, Schmidt hammer test, half-cell test, x-ray, ultrasound.
   b) destructive – compressive strength test, chloride ion test, removal of finished surfaces, core sample, soil sample.
4) Compressive strength testing of concrete involves drilling out a core sample and squashing it in a machine. You may want this done to confirm an observed potential problem or if there is a change in use of the building and you want to determine if the concrete can handle the new loads that will be imposed on it.
5) Soil testing tells you the bearing capacity of the soil and the compaction level.